This revised edition dedicated to the sense of humor and spirit of

Scott D. Anderson

Special Thanks to

Dr. Richard Anderson, U.S. Environmental Protection Agency entomologist, for verifying the accuracy of our information;

Dr. Paul Anderson, biochemist, for the dinosaur DNA data;

and Dr. Mary Sue Lux, veterinarian, for her help on questions concerning nonhuman mosquito victims.

No mosquitoes were harmed in the production of this book. (Sorry.) After you buy it, however, feel free to use it to whack the bejeebers out of the nasty little beasts.

Contents

Mosquito Life: Sex, blood, and standing water 1

Mosquito Facts: Interesting info about skeeters 33

Mosquitoes & Disease: History and facts 55

Controlling Mosquitoes: What works, what doesn't, and why 83

Repelling Mosquitoes: What works, what doesn't, and why 101

Mosquito Miscellany: Tall tales and other tidbits 127*

Bibliography: Where we found our facts 145

Index: A fast way to find our facts 149

**Additional "Mosquito Miscellany" has been disbursed randomly throughout this book. You'll know you've stumbled across it when you see this happy little fellow at the left. Those of you partial to drinking games can take a shot every time you see him.*

You should read this

We're from Minnesota, a place so thick with mosquitoes that many people consider the tiny pests the state bird. We are by no means, however, experts on mosquitoes. We're just some guys who've spent too much time waving our hands in the air trying to shoo away the annoying little bloodsuckers. We hate 'em.

Our hatred of mosquitoes prompted us to act. After years of blindly lashing out against the mini-blood banks that have made our summers unbearable, we decided to find out as much as we could about mosquitoes in order to deal with them as effectively as possible. What we found changed our lives, and we think it will do the same for you. That's why we wrote it all out in this book.

If, however, you use the information in this book incorrectly and somehow injure yourself or your loved ones, don't come to us with your fancy pants lawyers and their high-falutin' lawsuits. Even though we found a real card-carrying bug expert to verify our mosquito facts, this book is not a scientific text; it's just a fun little book filled with interesting stuff about mosquitoes. We hope you enjoy it.*

— The Authors

**Oh, and don't forget to check out the flip-book cartoon, starting on page 21.*

Mosquito Life:

Sex, blood, and standing water

What, exactly, is a mosquito?

According to *Merriam Webster's Collegiate Dictionary*, a mosquito is "any of a family (*Culicidae*) of dipterian flies with females that have a set of slender organs in the proboscis adapted to puncture the skin of animals and to suck their blood."

You know those bugs that look like huge, slow-flying mosquitoes, but never try to bite you? They're not mosquitoes! They're crane flies, of the family *Tipulidae*, and the adults don't eat (they're the supermodels of flies). Some people call them daddy-long-legs.*

**Of course, "daddy-long-legs" is also used to describe some big-ol' spiders. (Of course, daddy-long-legs aren't really spiders. They are arachnids, like spiders, but a separate order, so they're more like spiders' cousins—the embarrassing kind that you pretend not to see at the grocery store.)*

Mosquitoes don't need your blood for food.

Contrary to popular belief, mosquitoes don't get their nutrition from your blood.

Their nutritional needs are primarily met by flower nectar.*

**And, like many of us, the occasional microwave burrito.*

So why do mosquitoes need my blood?

Blood provides the protein mosquito eggs need for development. While fertilization occurs only after a blood meal, the blood does not fertilize the eggs.*

According to mosquito researchers Andrew Spielman and Michael D'Antonio, scientists don't know what part of the blood mosquitoes need for their eggs, how they differentiate species among potential victims, or how they produce sufficient sucking action to draw blood in through such a narrow tube. Perhaps it's just out of spite.

**As with humans, the male is necessary for at least one thing. That, and grilling.*

Is it true that only female mosquitoes "bite"?

Yes, what you've all heard is true: lady mosquitoes are the bloodsuckers of the family.

We will now pause to allow divorced men to make up their own joke.*

**Divorced women: yes, we realize he was only good for one thing—grilling.*

J. Edgar Mosquito

We just told you that only the female mosquito sucks blood, and we're going to tell you that throughout the book. And we're right. But here's the deal: there have been infrequent accounts of male mosquitoes sucking blood.

These very rare males, which have female markings to go along with their womanly behavior, are called "gynandromorphs," a term which refers to any animal exhibiting characteristics of both sexes.

Although these male mosquitoes suck blood, they are not able to produce eggs, so have no real need for feeding on humans.*

**Other than the chance to be just as annoying as the other half of the mosquito population.*

Mosquitoes don't actually "suck" your blood.

They pump it!

When we suck something through a straw, we use our lungs and cheeks to produce the suction to get the liquid. But the female mosquito has a specialized pump built into her head that she uses to pull in the blood. It works a lot like a turkey baster.*

**Since a mosquito's pump technically uses suction, we're still going to refer to the little bloodsuckers as "the little bloodsuckers."*

How do mosquitoes actually "bite"?

They don't bite: It's more like they go in and rip things up a bit. The female mosquito's mouth is composed of six long piercing parts called stylets. Once inserted into the skin (1), four of the stylets—armed with serrated edges—saw back and forth, ripping apart tiny capillaries just beneath the skin's surface to create a pool of blood (2). The fifth stylet injects saliva, which acts as an anticoagulant to help the blood flow smoothly (3). The last stylet is shaped like a trough. When the others wrap around it, they form a tube through which the blood is pumped (4). The mosquito then enjoys a little drink of blood (5).*

**If the mosquito is lucky, she will hit a vein and pump from there. If she's unlucky, she's bitten Keith Richards.*

How mosquitoes bite

(See text on facing page.)

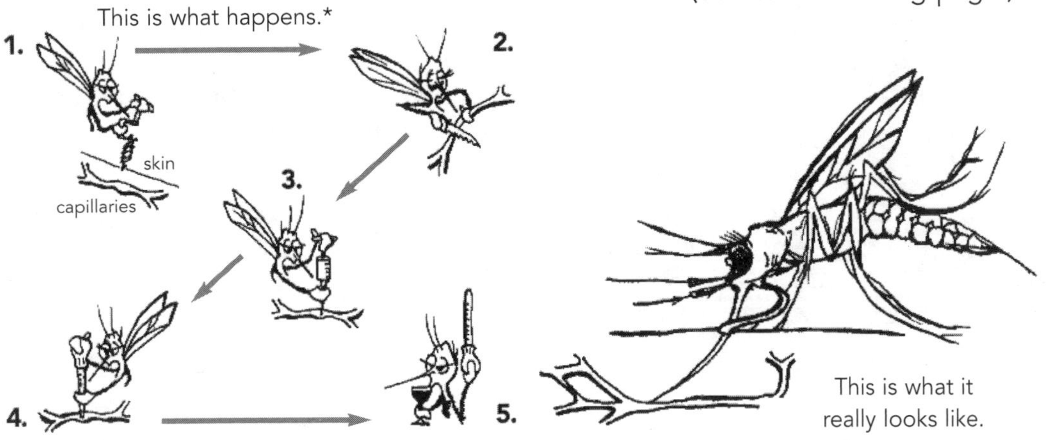

**Actual mosquitoes use no hand or power tools.*

It's a little like the *Alien* movies...

When the bee-sized female botfly of Central and South America is ready to lay her eggs, you might as well cue the horror movie music. The botfly swoops down on a female mosquito in midflight, grabs on, and sticks her eggs to the mosquito's abdomen.

When that mosquito eventually lands to feed, her unwanted passengers spring into action, stimulated to hatch by the body heat of the mosquito's victim. The maggots burrow into the flesh and continue to mature for twenty to sixty days, depending on the species, approaching two inches in length. When mature, they burrow back out and drop to the ground.

Can humans become infested with botfly larvae? Oh, you betcha.* Just look for the telltale breathing tube sticking out of a sore spot that you thought was just a mosquito bite. Blocking the tube often brings the larva wriggling to the surface; grab the little worm and pull it out.

**Just another one of the many joys that mosquitoes bring us.*

How much blood is lost from a mosquito "bite"?

The average mosquito consumes one millionth of a gallon of blood per "bite."

At that rate, it would take about 1,120,000 bites to drain the blood from an average adult human.

But that could never happen.*

**Because you'd be dead after losing half your blood—about 560,000 bites.*

How often can a mosquito bite?

Some people think mosquitoes, like bees, sting once and die. But remember, mosquitoes are developing eggs, not defending their queen. If they died after "stinging" once, it would defeat the purpose of taking the blood in the first place.

A female mosquito goes out for blood whenever she needs protein for her eggs. She can feed multiple times and, on average, makes between one and three batches of eggs during her lifetime (see pages 23 and 32).*

**Unless, of course, she gets squashed on her first attempt. Or has bitten Keith Richards.*

If bitten by a radioactive mosquito, do you acquire superpowers that allow you to fight crime and evil?

Yes, of course you do.*

The problem is, supervillians can hear the buzzing of your giant mosquito wings from up to a mile away. And they all carry DEET (see pages 102 to 104).

Well, not really. And you should probably have that bite looked at.

Do mosquitoes take blood only from humans?

No!

They'll take blood from just about anything that has it: mammals, birds, reptiles, even lawyers. According to Will Barker, author of *Familiar Insects of North America*, the female mosquito can puncture many types of body covering—even the leathery skin of a frog or the overlapping scales on a snake.* No wonder they can go right through clothes!

Lucky for us, the majority of mosquito species do not feed on humans, and of those many often feed from other sources.

**Tests on the skin of George Hamilton have so far been inconclusive.*

They eat their own!

While mosquitoes don't "bite" other mosquitoes, they sometimes go cannibalistic if crowded during the larval stage, and larger, older larvae might eat the smaller, freshly hatched larvae.*

Some mosquito species even prey on other mosquito species. In fact, scientists are working on methods to produce cannibalistic mosquito larvae as a population control method (for mosquitoes, that is).

**This probably sounds all-too familiar to those of you with older siblings.*

Can mosquitoes "eat" too much?

Not if everything's working right. If a mosquito gets too bloated with blood to fly away from her victim, she releases a little ballast to help her become airborne.

How?

She empties the mosquito equivalent to a bladder.*

**In other words, she piddles on you. Icky.*

How do mosquitoes know when to stop feeding?

The female stops feeding once stretch receptors in her abdomen have been triggered.

Experiments have shown that if the nerves connecting the stretch receptors to the brain are cut, the female will take blood until she bursts!*

**Kind of like when you undo your pants before having seconds at Thanksgiving.*

Turning the tables (or "Fun with skeeters!")

The next time a mosquito lands on your forearm, don't swat it. Instead, pinch your skin on either side so that the pressure traps her stinger in your arm—but not so tight that you cut off the bloodstream. Even when her sensors tell her it's time to stop, she'll keep taking in blood—until she explodes!*

**We've never actually tried this. It may be a myth...or is it?*

How do mosquitoes attract their mates?

Male mosquitoes—swarms of them—are attracted to a female by the whine given off by her beating wings.* Females' wings beat slower, and have a lower pitch, than males'. (The average wing speed of mosquitoes varies between 250 and 500 beats per second.)

This was unexpectedly demonstrated at a frequently malfunctioning Canadian power station. Engineers found thousands of male mosquitoes fouling the machinery. Apparently the hum of the station was at just the right frequency, leading to a painful end for those unsuspecting guys looking for love in the wrong place.

Scientists often use tuning forks set at the pitch of a female to attract (and ensnare) males.

**If you're not a mosquito, whining probably isn't the best way to attract a mate.*

Do mosquitoes have a mating ritual?

No. Mosquitoes mate after the female flies into a swarm of males—a huge mosquito singles bar if you will. This swarm may be as small as a softball or as large as a classroom, and mating takes place almost immediately—in midair! The happy couple eventually floats to the ground.*

**Mating takes anywhere from four to forty seconds (no giggling, ladies), but some couples have been known to stay together for over an hour (especially if listening to Barry White).*

Born backwards

After emerging from the pupal skin, the male mosquito is not attracted to females. A good thing, too: when a male mosquito emerges, his sex organs are on the wrong side of his body, making mating impossible (and buying a pair of pants extremely challenging).

This state lasts for about a day (depending on the species and the temperature). The terminal segments of the abdomen that hold the sex organs then rotate 180 degrees, and he's good to go.

Check out the flip-book cartoon. It starts here! →

Strange but painfully true!

In scientific experiments, male mosquitoes continued to copulate even after being decapitated!*

Strange but painfully true, too!

In other cases, the happy couple became so "attached" that the male was only able to move on by leaving his sexual organs behind (we've all had that feeling before).

**If you ask us, some scientists have way too much free time on their hands.*

How many eggs does a mosquito lay per "batch"?

Lots! Depending on the species and how much blood she has consumed, a female mosquito can lay as many as several hundred eggs in one batch.*

How many batches can the female produce?

The average female will lay one to three batches, but a long-lived mother mosquito could lay up to ten (see page 32).

Baby name books make great gifts for the expectant skeeter.

Does a female mate with a different male for each batch of eggs she lays?

No, she actually needs to mate only once before laying many batches, because she stores the sperm in her body—her own personal sperm bank—until she lays the eggs, making withdrawals to fertilize the eggs as needed. In fact, after she has the sperm she needs to reproduce, a female will fight off further attempts at mating.*

Male mosquitoes actually leave behind a pheromone during mating that makes the female less receptive to other males. While the average female mosquito will mate just the one time, males will go on to another six or so encounters.**

**Guys, insert your own sexist comment here. **Ladies, insert your own sexist comment here.*

Where do mosquitoes lay their eggs?

Many mosquitoes lay their eggs in standing water. Floodwater mosquitoes (*Aedes*) lay eggs at the edge of water; when the water recedes, the eggs lie dormant until the water rises again.

The *Anopheles* lay their eggs "loose" on the water surface, while the *Culex* lay eggs in "rafts" of about 250 connected eggs; eggs of both species hatch a few days after being laid.

In India, the malaria-carrying *Anopheles culicifaces* lays eggs while in flight!*

**Kinda like biological ordnance delivered by diminutive dive bombers.*

Abnormal nurseries

Cattail mosquito larvae are attached to cattail stems with a serrated syphon, which draws air from inside the plant, reducing predation because they don't have to get air from the surface.

And if you think that's a strange place to raise a kid, the pitcher plant mosquito lays her eggs in the purple pitcher plant, a carnivorous plant that eats insects, spiders, and sometimes even small frogs.*

**Also known as the* Fear Factor *Diet.*

Do mosquitoes have a family life?

Not really. As we said, mating is a one-night stand kind of affair. Although some tropical species defend their eggs to ward off egg-eating insects, they hardly get together for reunion picnics.*

**Unless, of course, it's to feed on human picnickers.*

Do mosquitoes migrate?

No—at least not intentionally. Some ride the wind until they are miles from where they were born.

Also, humans often unwittingly transfer species of mosquitoes across continents.

The Asian tiger mosquito was brought to North America from a scrap tire pile in either Nagasaki or Kobe, Japan, to Houston, Texas.*

**Today, the Asian tiger mosquito is found in twenty-one states.*

They're homebodies...mostly

Some species have been found 100 miles from where they were born, but most travel no farther than a mile or two from their birthplace, not unlike the Amish.*

Since they only fly about 2.5 m.p.h., who can blame them for not taking long trips?

Others just live in their parents' basements and play video games.

If they don't migrate, where do mosquitoes spend the winter?

Some adult mosquitoes in northern climates (if they don't first die of cold) spend the winters hiding out in barns, caves, tree holes, cellars, etc.—anywhere they can get out of the wind. To keep from freezing, they form glycerol, which acts as antifreeze. Most species overwinter as eggs, lying dormant until hatching conditions are right.

A few experts hypothesize, however, that mosquitoes simply spend their winters at home watching TV and are only out in summer because of reruns.*

**Other researchers have deemed this theory "just plain silly."*

What do mosquitoes do all day?

Most spend their day hiding in dark, cool places such as building cracks, tree holes, or leafy bushes.*

They only feed for an hour or so a day, but it may seem much longer because of the quantity of mosquitoes that lives around you. Different species eat at different times, but most feed during the last two hours of sunlight.

*During that time, many watch soaps or pursue hobbies such as knitting or model railroading.

How long do mosquitoes live?

Mosquitoes live an average of two weeks in the summer, but may live to be much older, depending on the type. *Culex* mosquitoes might reach a month in cold weather. *Anopheles* mosquitoes can live for several weeks (if they exercise and eat right). *Aedes* mosquitoes are mosquito Methuselahs—in cooler weather they can live five or six months.* Mosquito eggs can lie dormant for up to seven years waiting for ideal hatching conditions.

**In their later weeks, they like to play bingo and complain about popular music.*

Mosquito Facts:

Interesting info about skeeters

Who came up with the name "mosquito"?

In medieval England, a mosquito was called a "midge"—the word "mosquito" didn't appear until the sixteenth century. The etymology* of the word can be traced back to *musca*, the Latin word for "fly." This Latin word became *mosca* in Spanish and Portuguese. *Mosca* then became "little fly" or *mosquito* (also *mosquita*) to describe our favorite little pest. *Mosquito* was borrowed by English in about 1583.

Interestingly, the English word "musket" is also borrowed indirectly from Latin. The Latin *musca* and Italian *mosca* formed *moschetta*, meaning "bolt for a catapult" and "small artillery piece."

From *moschetta* came *moschetto*, or "musket," the source of French *mousquet*. And so "little fly" came to mean "bolt from a crossbow," appropriate enough since both crossbow projectiles and mosquitoes fly, buzz, and draw blood.

**"Etymology" (the study of words) and "entomology" (the study of insects) are in no way derivatives of one another. Linguists, however, often use them in elaborate puns to help their spouses fall asleep: "A guy who studies the roots of insect names walks into a bar...."*

The English call mosquitoes "gnats."

Of course, they also call car trunks "boots," elevators "lifts," and trucks "lorries." Besides that, they drink their beer warm.

One French word for "mosquito" can also mean "cousin."

French: the perfect language for those who consider their relatives annoying little bloodsuckers.*

**Those annoying little French have a different word for everything!*

A mosquito's brain is the size of the period at the end of this sentence. ←—!

And yet they have outwitted humankind since the dawn of history. Of course, they had a lot of practice before we came along. Mosquitoes have been consuming blood for about 200 million years.*

**Scientists have described the total volume of blood taken by mosquitoes in all that time as, "Holy Crap! That's a boatload of blood!"*

With such tiny brains, how do mosquitoes think?

Mosquitoes don't think, at least not in the conventional sense. Mosquitoes behave according to a set of fixed patterns that have been ingrained into their nervous systems after millions of years of evolution.*

Kind of like frat boys, but a bit more advanced, and without all the beer.

How much does a mosquito weigh?

A typical mosquito weighs around 2.5 milligrams.

That's about twenty thousand mosquitoes to a pound.*

**Thus the reason recipes featuring mosquitoes have never become popular: too darn much prep work. And the boneless, skinless variety is SO expensive.*

Mosquitoes prey on us, but what preys on mosquitoes?

Bats, birds, other insects, bacteria, fungi, lizards, spiders, fish, and (slap!) humans.

However, even though millions of mosquitoes become meals every day, predators do not significantly reduce mosquito populations.*

**This is why buying a purple martin or bat house as mosquito control is a bad investment (see pages 88 and 90).*

How many mosquitoes are out there?

About 100 trillion mosquitoes from 3,450 different species* are ready to pester humans on any given day.

Assuming the bugs are 1/4" by 1/4" by 1/4", 100 trillion mosquitoes stacked together on a football field would create a pile over three miles high (if they didn't squash each other).

**Fortunately, the U.S. hosts only 170 species; Canada has roughly 70 more, most of which are darn good hockey players.*

Where are mosquitoes found?

Pretty much everywhere.

Tropical regions have the most varieties of mosquito species, but the Arctic and Antarctic regions are home to larger populations of mosquitoes.

They have been found at elevations of 14,000 feet in the mountains of Kashmir and 3,800 feet below sea level in gold mines in Southern India.*

**And wherever you happen to be wearing shorts.*

9,000 bites per minute!

In Canada, during an experiment when hordes of mosquitoes actually darkened the sky (some mosquito swarms in the Arctic are the size of small states!), researchers who allowed themselves to be exposed to mosquitoes were bitten about 9,000 times in one minute.*

At that rate, adult human victims would lose half of their blood in about an hour—enough to kill them!

**Nearly 1,500 of these bites couldn't be scratched in public.*

Why do some mosquitoes buzz louder than others?

You may notice that mosquitoes seem to get louder when you are trying to sleep. They're not really louder: they're buzzing closer to your ears than they normally would because your head is the only thing not covered when sleeping.*

This same theory can be applied to spousal snoring.

Why is it so hard to swat a mosquito?

They may seem slow, but mosquitoes are tricky flyers: They can move up, down, sideways, or backwards.* However, it is most likely that the wind created by your swatting hand is actually blowing the mosquitoes away and keeping you from swatting them successfully.

**Some people claim mosquitoes can maneuver around raindrops, but this has not been proven.*

How do mosquitoes find us?

Mosquitoes live in a chemical world—they change their flight pattern depending on what they smell. Mosquitoes come a-callin' after bumping into chemical or physical clues coming from their potential victim: carbon dioxide, lactic acid, natural skin oils, and heat.*

**Basically, if you can prevent yourself from breathing and sweating, they'll leave you alone.*

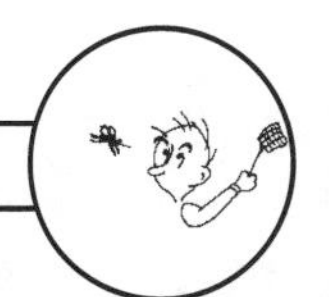

Why do mosquitoes seem to like humans so much?

Actually, we are the second choice for many species; some prefer birds.

We're also not the best meal for mosquitoes. Human blood is low in isoleucine, an amino acid mosquitoes need in order to build their egg proteins.

We are, however, easy prey—we're big and smelly, and there are many of us to bite.*

**The more we build (strip malls, housing, etc.), the more we force other animals to move, so we give the mosquitoes little choice but to chew on us.*

Why do mosquitoes go after some people more than others?

Some people are more attractive to mosquitoes because they smell better than others (or worse, depending on your perspective).*

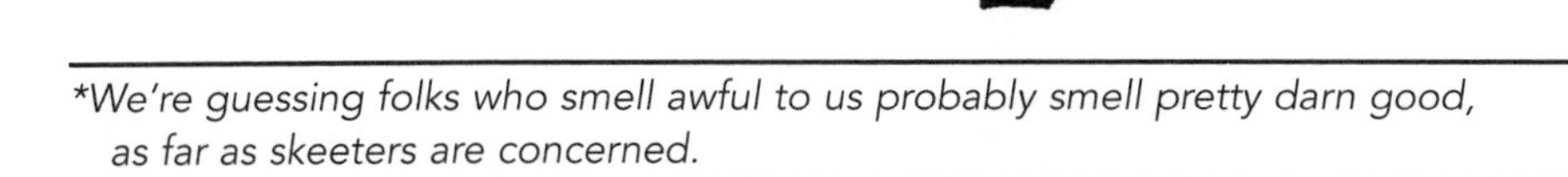

We're guessing folks who smell awful to us probably smell pretty darn good, as far as skeeters are concerned.

Why do mosquitoes like ankles so much?

We don't know, but our researcher, Laura Witrak, has a theory: "In the summer you wear sandals; your feet stink a little more (which attracts mosquitoes), and the ankles are near your feet. Voila—that's my explanation."*

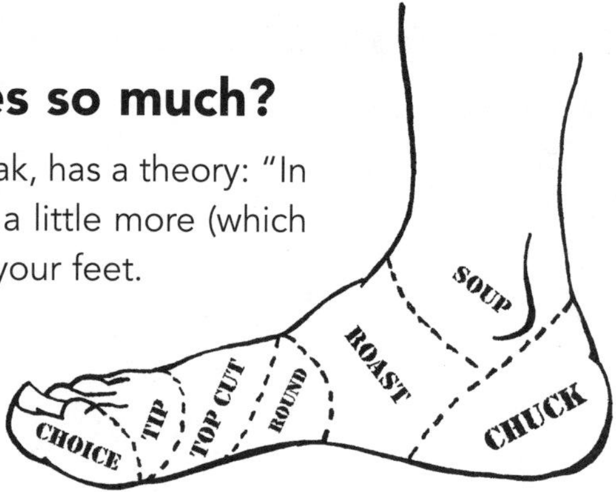

Speak for your own feet, Laura.

Why do you get "bumps" after a mosquito bite?

Before sucking your blood, the female mosquito injects you with her saliva,* which contains an anticoagulant that allows the blood to flow freely into her (this is also how she transmits disease; see pages 56–83). The "bump" is your body's reaction to a protein contained in the saliva.

Essentially, she spits in you. Icky.

Why doesn't everyone react to mosquito bites the same way?

Because it all depends on your skin sensitivity and predisposition to allergic reactions. It also depends on how many times you've been bitten in your lifetime. Some researchers have been bitten so many times they no longer react to mosquito saliva.*

**Just like those who listen to the radio so much they are no longer annoyed by Celine Dion.*

Poison!

You may have heard this before: "When you get bit, mosquitoes inject you with a poison—that's why you get an itchy bump. So wait until it's done feeding (and has sucked up the poison with your blood) before killing it."

Not true! The saliva contains no venom-like poison, just the anticoagulant* (and perhaps a virus; see page 49). Regardless, not all of what a mosquito injects gets sucked back out; if it did, mosquitoes wouldn't spread so many diseases.

So there's no need to wait before staking the little vampires.

Are some people *naturally* immune to mosquito bites?

Nope. People who have never been bitten by a mosquito will not react—except, perhaps, with a slap—to their first mosquito bite, but they are not immune.*

**It takes repeated bites before the skin reacts.*

Do nonhuman mosquito-bite victims itch like we do?

Yes! Horses, cattle, the family dog—mosquitoes make everything itch!*

We humans, however, seem to be the only ones complaining about it.

Are mosquitoes good for anything?

Many species of mosquitoes do help pollinate flowers. And they are a food source for other insects and organisms that don't bother humans. They are not, however, a significant pollinator, like bees, or a primary food source for insect-eating animals.

They can also help diagnose disease. Through a procedure called xenodiagnosis, physicians can diagnose an infectious disease at an early stage by exposing a presumably infected individual or tissue to a clean, laboratory-bred mosquito and then examining the mosquito for the presence of the infective microorganism.

Of course, they also spread disease, which helps keep populations of animals—and humans—under control (see the next chapter).*

**Overall, the mosquito seems to do only one thing really well: make more mosquitoes.*

Mosquitoes & Disease:

History and facts

Skeeters & disease: A biting history

Because of the diseases they spread, mosquitoes kill more humans than do any other animal. Even as humans began to evolve in Africa they were probably swatting disease-carrying mosquitoes. About 597 B.C., Babylon was besieged by plagues. Historians suspect that camel caravans—and the malaria-carrying mosquitoes that bred in irrigation canals along the caravan route—were the cause. Men in ancient China often planned the remarriage of their wives before traveling to areas known for malaria. The disease may have even helped prevent Genghis Khan from trying to conquer Western Europe. Columbus apparently brought malaria to the New World; there were serious epidemics in the West Indies in 1493.* Read on to learn more about how mosquito-borne disease has shaped our world.

"Ciao. You will mine gold for me, and in return I have brought enslavement and infectious disease."

Malaria

Malaria affects 300 million people per year, and it is transferred to humans solely by mosquitoes. A disease of the tropical world, malaria is found in 102 countries. Only a few cases have been found in the United States since the 1950s. Symptoms include high fever, chills, headache, fatigue, nausea, vomiting, diarrhea, anemia, spleen enlargement, liver and kidney failure, and finally brain damage, which can lead to death. Malaria has no vaccine. Quinine (the same stuff that's in tonic water) was once used to cure malaria, but the disease has developed resistance to it.* In the 1960s, malaria was thought to have been wiped off the face of the earth. Insecticides were killing mosquitoes and drugs were curing the last cases. The mosquito has since developed resistance to insecticides and drugs, and malaria is back, baffling scientists about how to stop it.

**Thus, "malaria prevention" is no longer a good excuse to drink gin & tonics. In fact, excessive G&Ts lead to some malaria-like symptoms, such as nausea, liver failure, and brain damage.*

The fall of Rome

Ancient Romans honored a "fever goddess" with a temple on the Palatine Hill, hoping for mercy from the deaths that came each summer. In Scotland, malaria took the lives of over forty thousand Roman soldiers. The lowland plain surrounding Rome was so malarious that many popes avoided the city altogether.

By A.D. 395, 330,000 acres of farmland in Rome's Campania region were abandoned due in part to a malaria epidemic brought on by mosquitoes that bred in nearby swampy areas. Rome fell just eighty-one years later.* (Invaders were unable to take full advantage of the fall of the Empire; armies of Visigoths, Ostrogoths, Huns, and Vandals were devastated by malaria, which also defeated Henry II's army in 1022.)

**Coincidence? You be the judge.*

Germ warfare, the old-fashioned way

When humans started moving long distances to colonize or invade, germ warfare happened naturally. Both sides only had immunities to their native diseases. Germs brought by newcomers are called offensive; the local infections are called defensive. Europeans brought offensive pathogens, such as small pox and measles, to the New World, decimating local populations; there were few defensive pathogens to help protect the natives. Europeans eventually received an ironic comeuppance. The decimation of the natives forced the invaders to look elsewhere for slave labor, and slave ships from Africa brought malaria and yellow fever to the white settlers of the New World.*

*British encounters with the defensive germs of West Africa led to its nickname, "the white man's grave."

Foiling the Spanish

Early Spanish expeditions to the Americas led by Hernando De Soto felt the wrath of mosquitoes. Half of his men never made it off American soil because of mosquito-borne disease. Of course, the natives defending their territories took quite a toll on De Soto's men as well.*

De Soto, the first European to see the Mississippi, was eventually buried in the mighty river after dying of fever in 1542. It is uncertain what sort of fever it was, but it is possible that he also was a victim of mosquito-borne disease.

**De Soto himself took an arrow in the, er, well, bum during one battle.*

The war in the Pacific—against mosquitoes

General Douglas MacArthur was losing the war in the South Pacific to malaria. A full third of his soldiers were sick, and another third were recuperating. He only had the remaining third to fight World War II.

This was a job for...Malaria Moe and Anopheles Ann? These *Stars and Stripes* cartoon characters were part of the army's campaign to remind soldiers to take their Atabrine, an antimalarial drug. Between the drug and antimosquito squads fighting the bloodsuckers, malaria rates went down. Having healthier soldiers was one edge the U.S. military had over its foes.*

**After the war, Malaria Moe and Anopheles Ann unsuccessfully tried to get work in the comic strips "Pogo" and "Nancy & Sluggo," but their feeding habits proved problematic.*

Buggin' in the 'Nam

Throughout much of the Vietnam War, malaria took more soldiers off duty than actual war wounds. Malaria was increasingly drug resistant, forcing the U.S. Army to look for new antimalarial substances. Mefloquine and halofantrine were two of the Army's discoveries. They starved the malaria parasite by interfering with its feeding. However, the drugs had side effects. Severe psychological problems were sometimes reported, such as hallucinations, paranoia, and suicidal behavior.*

**We're just going to let readers mull over the implications of mixing those symptoms with the Vietnam War by themselves. Good night, and have a pleasant tomorrow.*

Paradise lost

Hawaii was once a true tropical paradise: mosquito free! Unfortunately, in 1827 a whaling ship arrived with stowaway bloodsuckers on board, and Hawaii soon became the unpopular wasteland that it is today. Well, maybe it isn't that bad for humans, but that's how you'd feel if you were a Hawaiian honeycreeper, a four- to eight-inch bird of two varieties that feeds either on nectar or on fruits, seeds, and insects. When Europeans arrived in Hawaii there were fifty species of the brightly-plumed honeycreepers; after mosquitoes arrived, providing a vector for transmitting diseases such as avian malaria from migratory birds to the native birds, the number of species plummeted to twenty-one.*

**And fourteen of those twenty-one are now endangered.*

Yellow Fever

Yellow fever is another virus-driven disease transmitted by our friend the mosquito. It is rampant in monkey communities* and has been transferred to humans as we continue to encroach on the wild. This virus affects 10,000 people per year and is found in twenty countries. The virus moved from Africa to the Americas with the slave trade in the 1500s, but has not been found in the United States since 1905. Symptoms include sudden-onset fever, headache, nausea, slowed pulse, reduced urine production, and low white-cell count. In advanced stages, victims bleed from the mouth and nose, vomit blood, become jaundiced, and form lesions in the liver, kidneys, and gastrointestinal tract. Fortunately, there is an effective vaccine.

**Yellow fever has wiped out entire populations of monkeys in South and Central America.*

The nation's capital—1793

In 1793 Philadelphia was still the seat of federal government and America's busiest port. During that summer, refugees from a slave rebellion on Hispaniola arrived by the thousands, bringing yellow fever with them. Soon people were dying in such numbers—over one hundred victims a day during the epidemic's peak—that gravediggers could not keep up. President Washington and family went to Mount Vernon. Thousands fled. Cities imposed quarantines against Philadelphians. In the end, ten percent of Philadelphia's population perished: 5,500 out of 55,000. Winter came, frost killed the mosquitoes, and the epidemic faded.* The city recovered, but its citizens feared ships arriving from the West Indies for many summers after.

**President Washington ventured alone on horseback into the abandoned city on November 10 and found it livable again.*

Purchasing Louisiana

By 1801, slave-revolt leader François Toussaint-Louverture controlled the entire island of Hispaniola. Although the island had become a French protectorate, he did not obey Napoleon Bonaparte, who sent 33,000 men to regain control in 1802. Although Toussaint-Louverture did surrender, by 1803 yellow fever had claimed 29,000 of the French soldiers. Without the men to reoccupy Louisiana—recently reclaimed from Spain—and looking ahead to war with Britain and other expensive French expansions, Napoleon needed cash. President Thomas Jefferson was offered the entirety of the Louisiana Territory. Thanks to malaria-laden mosquitoes, the size of United States territory doubled.*

**At the time, Jefferson was only hoping to purchase New Orleans. (No doubt part of his obsessive pursuit of a truly authentic gumbo recipe.)*

Mosquitoes help free slaves

Recently captured slaves bound for Cuba aboard the Spanish schooner *Amistad* revolted against their captors on July 2, 1839, killing the captain and other crewmembers. The ship was eventually seized by the U.S. Navy, and the recaptured slaves were tried in Connecticut. Although Cuba had "legal" slavery, importing slaves from Africa was illegal. The judge therefore ruled the Africans had been kidnapped and had been within their rights to escape by any means necessary.*

The rebellion had been made a little easier by mosquitoes. The *Amistad*'s crew had been infected with yellow fever. Most of the slaves, due to previous exposure, were immune.

**His decision was upheld by the Supreme Court during appeal, and thirty-five survivors returned to Africa with the help of private donations.*

Big trouble in the Big Easy

In the 1800s New Orleans suffered a dozen outbreaks of yellow fever.* The 1853 epidemic claimed 9,000 of the city's residents and 11,000 more in the lower Mississippi Valley. Fatalities reached a peak of two hundred per day. As in Philadelphia in 1793, bodies could not be buried fast enough. Occupied coffins were left in stacks above ground. The heat caused the decomposing bodies to swell—some to the point that the coffins they were in broke open. Some said the sinful ways of the city had brought divine retribution; others blamed slaves and foreigners. During one outbreak in Louisiana, five Italian immigrants were lynched after being accused of spreading yellow fever by socializing with African-Americans.

**In 1905 a U.S. Public Health Service anti-mosquito campaign ended a yellow fever epidemic that killed over 1,000 people in New Orleans. It was the last epidemic of yellow fever in the U.S.*

His name was Mudd

Dr. Samuel Mudd had simply fulfilled his Hippocratic Oath by treating John Wilkes Booth's broken leg, but the military tribunal prosecuting the conspirators of Abraham Lincoln's assassination didn't see it that way. Mudd was convicted and received life imprisonment with three others who also escaped the noose. They were sent to Fort Jefferson, the "American Devil's Island," in the Florida Keys. The mercilessly hot island was plagued by mosquitoes. During a yellow fever outbreak in 1867, the prison doctor succumbed and Dr. Mudd, although infected himself, stepped up to help, saving the lives of several people. This time around, fulfilling his oath earned him a pardon from President Johnson, and he was released in 1869.*

**In 1876, the still-controversial Mudd ran for the Maryland legislature—and won.*

The Memphis blues

In 1878 over fifteen percent of the population of Memphis succumbed to yellow fever—over 5,000 deaths in a city of 33,000. Residents who fled the city could find no refuge; other towns would not let them in. Armed mobs met trains from Memphis to keep passengers from disembarking. Many on foot died, if not from the fever, then from dehydration, starvation, or exposure when they were not allowed into towns. Memphis legislators spoke of razing the city, salting the earth, and starting over in another location.*

**Some property owners, fearing the plan would be implemented, sold their land for much less than it was worth.*

A Frenchman finds a parasite

Dr. Charles Laveran of the French Foreign Legion had been sent to Algeria to study malaria in 1878, the year of the Memphis outbreak. The disease was rampant in French outposts, and the government wanted it stopped. In November 1880, Laveran microscopically examined blood he had drawn from a feverish soldier. He discovered something moving around in the blood and was certain he'd found what caused malaria. He was the first to witness malaria parasites in a blood sample, an important step toward figuring out how malaria spread; Laveran was eventually awarded a Nobel Prize for his discovery.* Now someone just had to figure out how the parasites got into humans.

**Turns out the French are good for something other than soft cheeses and selling us Louisiana.*

America vs. Spain and Cuban mosquitoes

Shortly before the Spanish-American War at a conference on yellow fever, experts considered a resolution to invade Cuba and attack the disease at its perceived source. Obviously, there were other factors, including the sinking of the USS *Maine*, but after the 1898 war Cuban mosquitoes were in trouble. When eighty percent of American occupying forces succumbed to yellow fever, Walter Reed led a commission to eliminate yellow fever from Cuba. After Reed proved mosquitoes transmitted the disease, Army doctor Major W. C. Gorgas isolated infected people and then went after the mosquitoes. Soldiers in Havana destroyed or emptied any containers of water and oiled lakes to kill mosquito larvae. Yellow fever lost this battle in just a few months.*

**And Gorgas became a hero. See pages 73 and 74.*

A man, a plan, a canal: Panama

In 1904, Major W. C. Gorgas, fresh from victory in Cuba (see page 72), led a team of scientists in the fight to eradicate mosquitoes in Panama in order to complete the Panama Canal. France began building the canal in 1881; by 1898 they were only halfway and had lost almost 30,000 workers to yellow fever and malaria. The original leader of the project, Ferdinand de Lesseps (who had also overseen the building of the Suez Canal), and his son, Charles, were sent to prison for their failure.* Once America had control of Cuba, it was in position to complete the canal. Gorgas eliminated yellow fever in just two years and significantly lowered malaria rates. The canal opened in 1914, the same year Gorgas became surgeon general of the U.S. Army.

**Gustave Eiffel, the French engineer of tower fame, paid a large fine for his involvement.*

Heroes of the mosquito wars

Sir Patrick Manson (1844–1922) Englishman Manson first made the assertion that mosquitoes transmit malaria (1877).

Dr. Charles Laveran (1845–1922) A doctor in the French Foreign Legion, Laveran was the first to identify the malaria parasite (1880). He received the Nobel Prize in 1907.

Sir Ronald Ross (1857–1932) A British physician, Ross won a 1902 Nobel Prize for proving that mosquitoes transmit malaria.

Carlos Juan Finlay (1833–1915) This Cuban physician suggested in 1881 that the mosquito was the carrier of yellow fever and later specified the correct species, now known as *Aedes aegypti.*

Walter Reed (1851–1902) An American army surgeon, Reed proved Finlay's theory of mosquitoes as the carrier of yellow fever.*

Major W. C. Gorgas (1854–1920) Coordinated the elimination of skeeters in Cuba and Panama.

**He also helped out with malaria problems in Cuba and has a big ol' medical center named for him.*

Filariasis

Filariasis is caused by filarial nematodes transmitted by mosquitoes. This disease is found in sections of Asia and Africa where sanitation is poor. It affects 250 million people in fifty countries, but has not yet been found in the United States. Symptoms include recurrent fevers, swollen lymph nodes, and swollen limbs. As the disease progresses, the limbs and genitalia become increasingly swollen.*

When filarial nematodes are ingested by mosquitoes, the little worms chew their way out of the mosquito's gut and into its body cavity. The nematodes eventually leave through the mosquito's mouthparts, entering their next host through the hole made by the mosquito's bite—hence filariasis. The hole left in the mosquito's gut allows other pathogens to migrate within the mosquito's body, possibly to be passed along with the mosquito's next bite.

**More easily amused readers can insert their own joke here.*

Dengue

Dengue is a fever sickness caused by a mosquito-borne virus. This illness has been recorded in the Americas, Africa, and Asia. Worldwide, nearly one million people in fifty countries are afflicted with dengue. It has not been found in the United States since the 1920s, when 600,000 people in Texas were affected. Symptoms include high fever, bone and joint pain, intense headache, skin rash, small hemorrhages, nausea, vomiting, swollen glands, fatigue, and depression. The pain is so intense dengue is also known as "break bone" fever.*

Dengue hemorrhagic fever, a more serious form of the disease, is even scarier. Its ebola virus-like symptoms include difficulty breathing, excessive thirst, and bleeding from the nose, mouth, and gums. It has been found in Latin America.

**Not to be confused with the short-lived TV show* Break Dance Fever.

Yellow fever, filariasis, dengue: if they're rarely, if ever, found in North America, should we even worry?

Just because they have not yet found their way to our part of the world doesn't mean we're safe from these diseases—consider the outbreaks of West Nile Virus in the U.S. in the early 2000s. If global warming theories are correct and mean world temperatures rise just a few degrees, mosquitoes carrying these diseases could increase their range and spread into North America.* Also, there is a low probability that an airline commuter could contract a mosquito-borne disease elsewhere, fly to the U.S., and infect others.

**And we were worried about communists and killer bees.*

Encephalitis

The one mosquito-borne virus infectious to humans that has a stronghold in the United States is encephalitis. This virus is passed back and forth between mosquitoes and either birds, chipmunks, or squirrels until an infected mosquito bites an amplifying host (an animal that does not get sick from the virus but instead helps the virus to mature and reproduce). For example, a mosquito carrying immature encephalitis virus bites and infects a chipmunk. The virus matures in the chipmunk's blood. Later, another mosquito bites the same chipmunk and picks up the mature virus. The mosquito can now pass it on to humans, who experience horrible symptoms: sudden-onset fever, headache, stiff neck, brain inflammation, convulsions, and sometimes coma.*

**People who do survive these bouts become immune to the viruses.*

West Nile Virus

Commonly found in vertebrates in Africa, Eastern Europe, West Asia, and the Middle East, West Nile Virus (WNV) first appeared in the U.S. in 1999.* While it's sensible to take precautions, there's no reason to panic. Nearly eighty percent of people who are infected develop no illness. Twenty percent will suffer only from West Nile fever—flu-like symptoms (sometimes with a rash on the trunk of the body) that last just a few days.

Only about 1 in 150 cases develop more serious conditions that inflame the brain and spinal cord: West Nile encephalitis, West Nile meningitis, and West Nile meningoencephalitis. Their symptoms include headache, high fever, neck stiffness, stupor, disorientation, coma, tremors, convulsions, muscle weakness, and paralysis. The fatality rate among these infections is less than fifteen percent, so of all people infected with WNV, under 1 in 1,000 will die.

In addition to people, WNV infects birds, horses, and some other mammals. Cats and dogs have shown few symptoms even if infected. It can be fatal in horses, however.

WNV in the USA

In 1999 a sixty-year-old New York man with headaches and paralysis was diagnosed with encephalitis. Three more cases quickly followed. At the same time, hundreds of birds had died and dozens of horses were suffering from apparent loss of balance or inability to control their limbs. It was thought to be St. Louis encephalitis, but the number of species infected was unheard of. People panicked, spraying began, and outdoor events were cancelled. The U.S. Army Medical Research Institute of Infectious Diseases eventually identified WNV—the virus's first known appearance in the Western Hemisphere. Soon there were sixty-two confirmed human cases and seven deaths. In 2003, there were nearly 9,000 cases in the U.S., with just over 200 deaths.*

**Remember, for each confirmed case there are dozens of people infected without knowing it.*

Heartworm disease

Most dog owners are well aware that mosquitoes transmit heartworm disease to dogs. This disease, a parasitic worm that can grow up to fourteen inches long and lives in a dog's pulmonary arteries, is found all over North America. There is no vaccine, but monthly doses of antiparasitics called milbemycin oxime and ivermectin prevent dogs from developing the disease by killing the worm while it's still in its larval stage. If your dog is not taking a heartworm preventive medication, get that pup to the vet!*

Cats can get it too, but it's not as common in felines as it is in dogs.

Can mosquitoes transmit AIDS?

No!

While mosquitoes can spread more than one hundred viral diseases and naturally ingest the HIV virus whenever they take blood from an infected person, the AIDS-causing virus dies quickly inside mosquitoes' stomachs.*

**They can't carry the Ebola virus, either!*

Controlling Mosquitoes:
What works, what doesn't, and why

The dawn of mosquito control

Humans have been trying to control mosquito populations for a long time and have dumped everything from oil of pennyroyal to crude oil in mosquito breeding grounds in an effort to get rid of them. Unfortunately, the crude oil also killed every other species in the water and ruined entire ecosystems.*

Today, the U.S. and Canada spend about $150 million a year trying to control mosquito populations.

**This is what's known as throwing out the baby with the bath water. Or destroying a village to save it.*

Mosquito bodies in the Meadowlands

Beginning in 1900, New Jersey's state entomologist, John B. Smith, went to war with mosquitoes. After identifying the dominant mosquito of the Garden State, he focused on its favorite breeding grounds: the Meadowlands, a brackish swamp north of Newark (and home to the New York Giants). Drainage ditches were dug to lower the swampy water, and the mosquito population took a nose dive.*

Smith's work on the Meadowlands and other Jersey wetlands led to a sharp decrease in malaria.

**Mobsters rejoiced. "It took a long time to bury a body when you had to keep stopping to swat mosquitoes, you know? Now I get back while the pasta fagioli is still warm."*

Dragonflies

Dragonflies, along with damselflies, eat lots of mosquito larvae and adults. In fact, in the southern U.S., dragonflies are often called "mosquito hawks" or "skeeter hawks."*

**Pretentious, brooding thespian dragonflies are called "ethan hawks."*

The mosquito fish

Around the turn of the twentieth century, the mosquito fish (the most wide-spread freshwater fish in the world) was used to control mosquito populations. A little gray fish from the northeastern U.S., the mosquito fish is a rabid eater of mosquito larvae and can clear out an entire pond within days. But they're not perfect. You see, in ponds with dense floating vegetation and organic debris, mosquito fish ignore the larvae. They also consume other beneficial insect larvae and fish that eat mosquito larvae. And they tend to take over their new habitats and shove out other species.* For these and other reasons, this practice is not in widespread use today.

**Much like we do here in America.*

Purple martins

Although many people think that purple martins eat lots and lots of mosquitoes and that having them around will keep mosquito populations low, there is no biological evidence to support this claim.

Some experts believe this myth was spread by a certain "Dr. Wald" from Libertyville, Illinois, who claimed that a purple martin eats "6,600 mosquitoes a day." No one knows where Dr. Wald got his information, but they do know his profession. Dr. Wald sold aluminum bird houses designed specifically for—you guessed it—purple martins!*

**Today, Internet sites market purple martin houses for moth, flying ant, fly, and mosquito control.*

DDT

The pesticide DDT (dichlordiphenylethylene) was once quite effective at controlling mosquitoes. But since mosquitoes evolve at such a fast rate they developed a resistance to the chemical.* Evidence also surfaced that DDT killed fish and other animals, and that it weakened the shells of bird eggs—especially those of eagles. At the end of 1972, the Environmental Protection Agency banned the general use of DDT in the U.S. However, others claim that the negative environmental impact resulted from large-scale, nondiscriminate spraying of DDT, and that careful application targeting mosquitoes is much safer. They also argue that any remaining negative effects are justifiable compared to the one million people who die each year from malaria. It is a debate that continues to this day.

**Unfortunately, humans evolve at such a slow rate we can't seem to find anything to effectively control mosquitoes. Or stop the spread of so-called reality televison.*

Bats

Many people believe that because bats dine on insects, building a bat house on their property will keep mosquitoes away. They are wrong. While bats eat lots of insects, mosquitoes make up just a fraction of their diet. Bats mostly enjoy beetles, moths, and leafhoppers. A recent study showed that mosquitoes made up only 0.7 percent of a bat's stomach content. Basically, bats eat whatever is handy, so you can't expect them to focus on just mosquitoes. And bats tend to migrate south in late summer, so there aren't many of them around when mosquito populations are at their peak.

**One Internet site selling bat houses claims "One bat can catch 500 mosquitoes in an hour–that's fewer insects to bite you!"*

Bti

Bti is a biopesticide designed to destroy mosquito larvae. Biodegradable within forty-eight hours, it contains bacterial spores whose toxic crystals destroy larval stomach linings, killing mosquitoes before they develop enough to get off the ground.

The manufacturer of "Mosquito Dunks" claims placing just one of its Bti "donuts" in your property's standing water once a month will keep your yard mosquito-free. The naturally degradable donuts float on the water and slowly release the Bti.*

**Warning: not to be used as a floatation device; ineffective if dunked in coffee.*

Malathion and naled

Malathion and naled are both organophosphates (OP), insecticides used primarily for controlling adult mosquitoes, but also used on food and feed crops and in greenhouses. Manufacturers claim that when applied properly, these pesticides can be used to kill mosquitoes without posing unreasonable risks to human health or the environment. They are usually applied by truck-mounted or aircraft-mounted sprayers that dispense very fine aerosol droplets that stay aloft and kill mosquitoes on contact. They degrade rapidly in the environment and if used properly pose little risk to birds and mammals.*

**However, both are highly toxic to insects, including beneficial insects such as honeybees.*

Goldenbear oil, Altosid, and Scourge

Oil has been used against mosquitoes since the beginning of mosquito control, often with horrible side effects. Goldenbear is a very thin oil that avoids most of those problems.* When sprayed onto water it spreads out into a monomolecular film. It is enough to suffocate the larvae, but in a few days it evaporates. It cannot be used if there are fish in the water, however (for the sake of the fish!).

Altosid is a larvaecide which has little impact on non-target species. The insect growth regulator it contains prevents mosquito larvae from developing properly.

Scourge is an insecticide that targets adult mosquitoes. Sunlight breaks down the active ingredient, Resmethrin, in under four hours, limiting negative side effects.

Goldenbear oil is **not fluid produced by legendary golfer Jack "The Golden Bear" Nicklaus.*

The black cigars of death

In 1909, Brazilian rubber plantations stocked up on *Il Negro Mortes*, "the black cigars of death," in an effort to control mosquitoes. Workers earned extra cash smoking the cigars—made of tobacco, chemicals, and petroleum tar—on the verandas of plantation owners' mansions. The fumes were not only effective on mosquitoes, but deadly to all: an estimated one smoker died for every 5 billion mosquitoes killed.*

**Perhaps the first tobacco-related deaths due to chemical additives.*

Trap 'em!

Recent years have seen the advent of mosquito traps. They come in all shapes and sizes and run on a variety of power sources, but they work pretty much the same: by using propane to emit carbon dioxide that lures female mosquitoes to a capture or killing device.* Some of these traps capture mosquitoes using an impeller fan that draws them into a net, some use a sticky surface that prevent mosquitoes from flying off once they land, and others electrocute mosquitoes on contact.

While the devices do indeed work, claims of acre-wide control may be stretching it a bit. Success depends on many factors, including mosquito population, proximity, size, type of breeding habitat, and wind velocity and direction. Oh, and they just might attract more mosquitoes to your yard than they can actually trap.

*The CO_2 attracts blood-feeding insects, so insects like moths and beetles are usually safe.

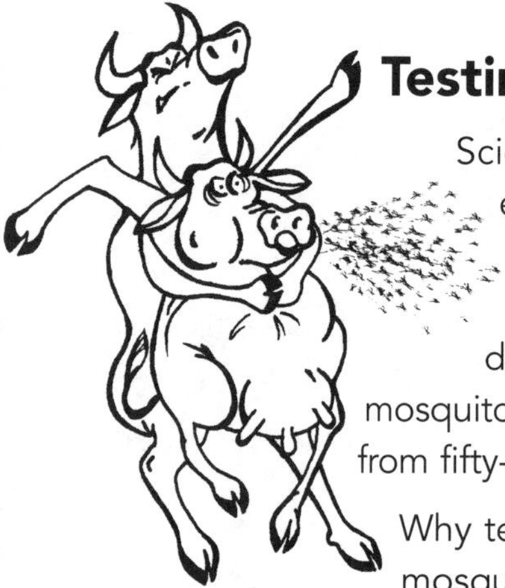

Testing the traps

Scientists used Florida's Key Island to stage the first experiments with mosquito traps. They essentially used large barrels baited with carbon dioxide and a liquid octenol solution (which smells like cows' breath) and packed in a dose of a synthetic pyrethroid insecticide that destroyed the mosquitoes on contact. After a one-month test, the mosquito carnage from fifty-two traps filled three thirty-quart coolers.*

Why test on Key Island? Because it's thick with skeeters. In 1989, mosquitoes were so thick they were killing the island's cows: the beleaguered bovines inhaled so many mosquitoes they were choking and suffocating.

**That's more than two billion mosquitoes!*

The future of mosquito control, part I

Today scientists continue to develop biocontrol methods.* One is a protozoan parasite (microsporidia) discovered in Thailand; it produces spores in adult female *Aedes aegypti* mosquitoes. The spores don't kill the female, but they do infect her eggs, killing the larvae before they can mature. As the dead larvae disintegrate, the spores are released, infecting any other larvae in the same water.

Scientists in Singapore are transplanting genes from *Bacillus sphaericus*, a natural insecticide, to a more hardy bacterium (*B. sphaericus* dies quickly under sunlight). They hope the bioengineered bug will become a hardier delivery system. At the University of Memphis, scientists have moved genes from a strain of *Bacillus thuringiensis*, another well-known insect killer, into a blue-green algae in an attempt to persuade the transgenic algae to produce higher amounts of toxin.

**In California, they're developing fungus-based pesticides to kill skeeters that breed in crops.*

The future of mosquito control, part II

The U.S. National Institutes of Health and the World Health Organization, among other such agencies, have started to focus on genetic engineering as a solution to mosquito-borne disease. The idea is to genetically alter some mosquitoes, render them incapable of carrying a disease, and then release them into the wild to breed and pass their modified genes along, eventually producing a wild population that is disease free.

This raises several questions. Will the engineered mosquitoes survive and breed in the wild? How do we keep the modified mosquitoes safe from other control methods so that they are able to pass along their genes? Will genetically-engineered mosquitoes become super intelligent and take over the world in a *Planet of the Mosquitoes** scenario? There are many problems to solve. The odds are for the mosquitoes.

**Charlton Heston says, "Take your stinking stylets off me, you damn dirty mosquito!"*

Back to the future of mosquito control

We have mentioned the possibility of genetically-modified mosquitoes being used to stop mosquito-borne disease. Actually, a biologically-based antimosquito campaign was already tried. The World Health Organization attempted to control the mosquito population in India by releasing male mosquitoes which had been sterilized by radiation. Since most females only mate once, mating with a sterile male would essentially sterilize them as well. It seemed like it could work.

However, the people in the test region somehow became convinced that they were the subjects of the sterilization program, and that it was being carried out by the CIA. The program was quickly terminated.

A poorly-informed, paranoid populace? Or a new vector for biological warfare?*

**Think about that the next time you lay awake in the dark hearing the tell-tale whine of wings. Sweet dreams.*

Yard and home mosquito control

Most mosquitoes breed in standing water, so do all you can to reduce standing water around your house (5,000 mosquitoes can come from a single tire filled with standing water!):

- Keep grass cut short and shrubbery well trimmed around the house (mosquitoes love to hide in tall, cool grass and the shade of shrubs).
- Use sand to fill in any holes in trees or hollow stumps that hold water.
- Fill in holes and low spots in your yard (where water can build up).
- Empty your child's plastic wading pool and store it indoors when not in use.
- Change the water in bird baths and plant pots often.
- Keep drains and ditches free of debris (so water will drain).
- Cover trash containers.
- Remove any old tires, buckets, or empty planters.
- Repair leaky garden-hose faucets or any pipes that lead outdoors.*

**If all else fails, move to a drought-stricken area.*

Repelling Mosquitoes

What works, what doesn't, and why

DEET: effective but dangerous

In the U.S., 50 to 100 million cans of insect repellent are sold each year, so some of it must work, right? Those containing the chemical DEET (N,N-diethyl-3-meta-tolumide) do, at least to some degree. DEET fools biting insects by masking our odors.

Repellents that contain DEET are by far the most effective repellents currently on the market. However, too much DEET can cause skin reactions, eye and sinus irritability, insomnia, headaches, and confusion. (It can even dissolve nylon and plastic!*)

If you suspect that you or your child is reacting to DEET or any insect repellent:

1. Discontinue use immediately. **2.** Wash DEET-treated skin with soap and water. **3.** Call your local poison control center. **4.** If you go to a doctor, take the repellent container with you so the doctor can identify the ingredients.

**If you don't like the idea of smearing DEET on your skin, try using a micro-encapsulated controlled release product that prevents DEET from entering your skin. Some claim to work up to twelve hours.*

Careful with that DEET!

- Do not apply over cuts, wounds, or irritated skin.
- Do not apply to hands or near eyes and mouth of children, and do not allow young children to apply DEET without adult supervision.
- Use just enough repellent to cover exposed skin.
- Do not use DEET under clothing.
- Do not spray in enclosed areas.
- To apply to face; spray on hands first and then rub on face. Do not spray directly onto face.
- Do not use sunscreens that contain DEET, because sunscreen needs to be applied often and liberally and DEET must be applied sparingly.*
- After returning indoors, wash treated skin with soap and water.
- Do not use DEET on your pets.
- Always read a repellent's label and follow its directions to the letter. Do not over apply (see next page).

When used together, DEET can reduce a sunscreen's SPF by a third.

Got DEET?

If you do, be careful to use the right amount. Check all repellent labels for the amount of DEET used in the product and follow theses recommendations:

Age	Concentration of DEET	Number of Applications
Children under six months	**Do not use any DEET**	**None**
Children age six months to two years	10% or less	Once per day
Children between two and twelve years old	10% or less	Up to three times a day
More than twelve years old.	30% or less*	Repeat as necessary according to label

*Most experts agree adults need more than 10% *only* if they are in areas extremely thick with skeeters.

Avon's Skin-so-Soft

Folks have sworn by Skin-so-Soft for years. It contains diisopropyl adapt benzophenome, which has a mosquito-repelling power, but is not as effective as DEET. Scientists consider it "unreliable." When tested under laboratory conditions, it provided thirty to forty minutes of protection from bites. Researchers believe the repellent effect could be caused by its fragrance in combination with individual body odors.

Avon has never claimed that Skin-So-Soft repels skeeters, but they now market a repellent called Bug Guard.*

**Now you big burly types have to come up with some other excuse to get baby-smooth skin.*

Plant oils

Lots of plants have had their oils extracted in the name of repelling skeeters, including allspice, basil, cedar, cinnamon, citronella (see next page), garlic, geranium, lavender, lemon eucalyptus, pennyroyal, peppermint, pine, rosemary, soybean, and thyme. However, according to the *Annals of American Medicine*, none of the plant-derived chemicals tested to date demonstrate the broad effectiveness and duration of DEET (none last more than two hours).

But it seems that the folks at Consep, Inc., of Bend, Oregon, have found a combination that works best. Their Bite Blocker contains soybean oil, geranium oil, and coconut oil—and no DEET. Studies conducted at the University of Guelph, Ontario, Canada, showed Bite Blocker provided more than ninety-seven percent protection against *Aedes* mosquitoes for up to three-and-a-half hours.*

**Unfortunately, not all skeeters are* Aedes.

Citronella

While some studies claim people receive forty-two percent fewer bites when burning a candle made with citronella oil (found in grass of China), these devices are most often ineffective. The candles give off chemicals that drive mosquitoes away, but only if there is little or no wind.*

Citronella sprays claim to protect people against mosquito for up to two hours. Studies indicate that citronella-based products can cause allergic reactions in some people. (Never use it on children under two years old.)

**Never light one of these bad boys indoors—the skeet-spookin' chemicals are also harmful to your health.*

The citronella and citrosa plants

If the oil works, so too must the plant, right? That's the marketing behind sales of the citronella plant. However, the plant must be squeezed frequently in order to release the skeeter-repelling oils. Tests have also indicated citronella plants offer no protection against bites.

The citrosa plant is a cross between the African geranium and grass of China (which contains citronella oil). Because the oils are emitted only when the plant is touched, the mosquito-repelling properties are rarely released. Studies do not support the idea that the citrosa plant actually repels mosquitoes.*

**Other plants considered mosquito deterrents include lemon thyme, lemon balm, citronella grass, tomatoes, sweet basil, marigolds, and ox-eye daisy.*

Geraniums and chrysanthemums

Venice, Italy, is infested with mosquitoes: all those canals mean lots of standing water. Almost everywhere you look, window boxes are filled with geraniums. Why? To ward off our buzzing buddies. Apparently geraniums harbor a citronella-like chemical. Research does not support geraniums as an effective repellent.

Using chrysanthemums as a mosquito deterrent makes much more sense than using geraniums: pyrethrum, a derivative of which is used in propane foggers (see page 115), is made from the extract of chrysanthemum. Insect repellents such as Pet & Premises are designed for dogs and cats and include chrysanthemum extract because it is safe for the animals and indoor use. They generally cause insects to have spasms which knock them to the floor. Then you can just vacuum them up!*

**Venetians have apparently turned a blind eye to this fact. (Sorry.)*

Folk repellents or myths?

Folks have sent us their favorite folk remedies, but do they really work? Try 'em and find out!*

- Cover your skin with Vick's Vaporub.
- Drink tea made of cat mint and rosemary.
- A clove-studded orange pomander will keep mosquitoes and moths at bay.
- Eating lots of garlic and spicy foods every day will help keep those nasty ol' skeeters away. (Scientists say, "Untrue!")
- Taking vitamin B1 (Thiamine Hydrochloride) will make you repellent to mosquitoes. (The FDA considers this claim "unsupported.")
- Chew on a match—the sulphur you sweat out will act as a repellent (used by U.S. troops in Iraq).
- Mix a little water and a few of drops of Lemon Fresh Joy dishwashing soap in a white dinner plate and set it outside. Folks claim mosquitoes flock to it and drop within ten feet of the plate.

**Warning: Many of these ideas will keep more than just mosquitoes away from you.*

Mosquitoes, monkeys, and millipedes

According to many Web sites (so it must be true), mosquitoes love bananas, or at least the smell a body gives off after that body has consumed bananas. And everyone knows monkeys love bananas. So what's a monkey to do to keep from being bitten?

Wedge-capped capuchin monkeys of Venezuela rub themselves with four-inch-long millipedes. No, it's not just for the fun of it. This particular species of millipede exudes defensive chemicals to protect itself from insects. By rubbing the millipedes on their fur, the monkeys are, in effect, using insect repellent.

To get the insect's toxic juices flowing, a capuchin monkey will frequently put a millipede in its mouth before proceeding to rub the bug on its body. The chemicals are so strong that just that short time in the mouth will cause the monkey to drool and its eyes to glaze over.*

**Humans pay good money for that sort of thing.*

Pyrethrum and permethrin

Pyrethrum is a fast-acting, powerful insecticide that was first developed from dried and crushed flowers of the daisy *Chrysanthemum cinerariifolium*. Permethrin is a human-made synthetic pyrethroid.

These chemicals don't repel insects by shooing them away with chemical signals. They are contact insecticides which cause nervous system toxicity that leads to the sudden death of mosquitoes called "knockdown" because they literally fall dead out of the air.*

**It is also effective against flies, ticks, and chiggers.*

Mosquito nets and clothing

Lots of companies sell mosquito netting and so-called mosquito suits, but these products are much more effective with a dose of permethrin. Luckily, a spray form is available. It is nearly odor free, won't stain fabric, and will maintain its potency for up to two weeks no matter the conditions, even after several washings.

And it works pretty good: In an Alaskan field trial, subjects wearing clothes treated with permethrin (as well as a strong dose of DEET on their exposed skin) were bitten on an average only once an hour for eight hours.

Careful, though: Follow the package's directions to make sure you apply it properly, and *never* apply Permathrin to your skin.*

**Or anyone else's, for that matter.*

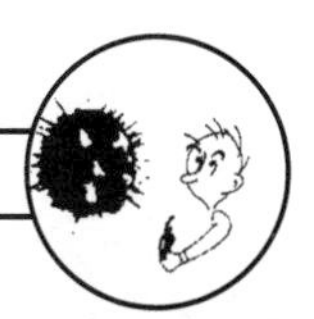

Mosquito coils

Mosquito coils are plant-derived devices that you set on fire. The smoke they give off contains mosquito-repelling chemicals. What's the chemical? Most contain pyrethroid, the same chemical found in foggers. Other uses citronella and are marketed as citronella incense.* They aren't nearly as effective as other products that employ the same chemicals.

**As with citronella candles, never light any mosquito coil indoors!*

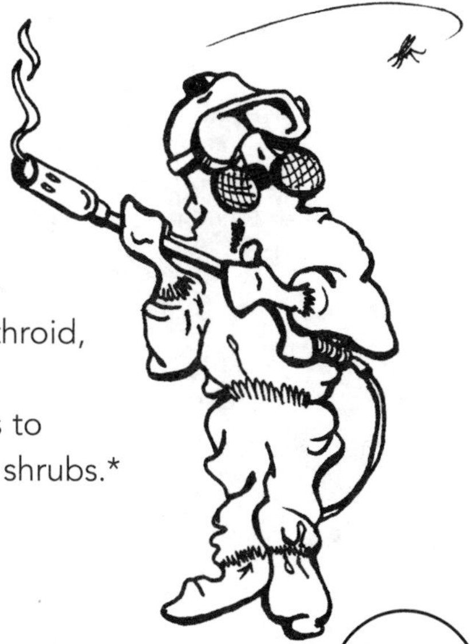

Propane foggers

These work well in small areas, but they are not cost efficient: they are expensive, and their effects are only temporary. The chemical used in these foggers is pyrethroid, a derivative of pyrethrum made from the extract of chrysanthemums. Like pyrethrum, it causes mosquitoes to have spastic attacks. It lasts longest if sprayed on lawn shrubs.*

**Don't even bother with foggers on windy days.*

"Bug zappers"

For about sixty dollars you can buy a bug zappers, machines consisting of ultraviolet lights that attract insects and electric coils that fry them. These machines do in fact attract and kill insects. However, mosquitoes compose less than seven percent of the catch, and only half of those are blood-feeding females. In fact, the zappers end up zapping a lot of insects that would have fed on more mosquitoes than the zapper can zap!* Zappers can really backfire on you: Mosquitoes are attracted to ultraviolet light, but they are even more attracted to your smell. So when you set up a zapper in your yard, you end up attracting mosquitoes that, once they catch a whiff of human, turn from the zapper to you!

**Over ninety percent of all insects killed in zappers are beneficial insects.*

Sonic devices

For a cost of anywhere from five to forty dollars, you can purchase electronic devices that emit a noise at a frequency designed to repel mosquitoes. They're even available as bracelets!

Careful with your cash—many scientists consider them a waste of money.*

In other words, sound devices are not effective enough to be considered sound investments.

Smudge pots or gardeners' "smokers"

Some gardeners swear by this technique: Soak a rag in motor oil, put the rag in a metal bucket placed near your chores, and burn, baby, burn. The fumes will keep skeeters at bay.

Of course, the toxic fumes might just keep you away as well. Not to mention the irony that the burning of fossil fuels may be contributing to global warming, and higher temperatures could lead to more mosquitoes.

This is your basic "penny wise, pound foolish" proposition.*

**And we don't advocate anything that requires you to set fires. Unless it's to roast marshmallows.*

Bounce fabric softener?

Florida residents have been spotted wearing Bounce fabric softener sheets on their belts. Apparently they believe the scent keeps skeeters at bay (there is no scientific evidence that the sheets actually keep mosquitoes away).

A fashion faux pax? Not necessarily. Many residents of Florida are retired gentlemen over age sixty-five: They have a tendency to wear white shoes and matching belts, so others rarely notice the fabric softener sheet.*

**Not on the AARP's mailing list? Tape a Bounce sheet to a fan to blow scent around room.*

Wallowing in mud

Elephants, while being pachyderms (literally, "thick skinned"), actually have quite sensitive skin, thick though it may be. So mosquitoes are, as always, a nuisance, even to the largest living land animal. In response, elephants often cover themselves in mud, which, in addition to helping them keep cool, also keeps the mosquitoes off them.

These are the same reasons their fellow pachyderms, pigs, are known for wallowing in the mud.

Unfortunately, there are few circumstances under which this technique can be used by people.*

**Except, perhaps, at large outdoor rock concerts with plenty of rain and dirt.*

The Beothuk: repelling skeeters and Vikings

Natives of Newfoundland, the Beothuk were apparently first encountered by Europeans when Vikings visited Newfoundland several times between 985 and 1011. The Vikings attempted to establish a colony in what they called Vinland, but natives repelled them. The last known Beothuk, Nancy Shawanahdit, died in 1829.

In June 1497, Genoese captain John Cabot, in the employ of Henry VII, landed in the Newfoundland area and claimed it for England. He eventually encountered the Beothuk, whom he called "red Indians" or "red skins" because of their habit of smearing themselves with red ochre for religious reasons and as a mosquito repellent.*

**Unfortunately for the Beothuk, the ochre did not repel this second wave of Europeans. The Europeans stayed, and the Beothuk were enslaved and hunted to oblivion.*

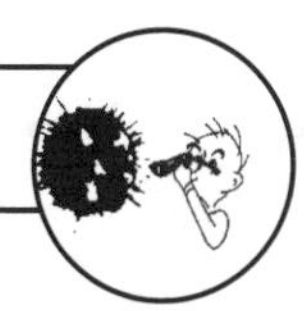

Native American repellents

According to outdoor writer Jeff Rennicke, Native Americans have traditionally used a variety of natural substances to deter mosquitoes, including deer fat, onion juice, cedar oil, and smoke smudge. Rennicke found that the best was cedar oil, which smells wonderful and left him mosquito-free. The most popular Native American method for dealing with mosquitoes is also the simplest: ignore them.*

**This is harder than it sounds: ignoring mosquitoes won't make them go away.*

Personal mosquito evasion tactics

Want to do as much as you can to avoid mosquitoes? Follow these common sense defense mechanisms:

- Wear loose clothing (the looseness provides a pocket of protection around you).
- Wear white or drab-colored clothing (mosquitoes are less attracted to dull colors). Never wear red—you might as well be ringing their dinner bell.
- Tuck pants into socks (so mosquitoes can't fly up your pants legs).
- Cover your head (wear a hat or buy a shirt with a hood).
- Stay inside during dusk and dawn (when mosquito feeding is at its peak).
- Avoid eating bananas.
- Do not wear a lot of smelly lotions, perfume, or hairsprays (remember, mosquitoes are attracted to certain smells).
- Smoke from a campfire will keep mosquitoes away from the immediate area.*

**If all else fails, wrap yourself in a huge plastic zipper bag.*

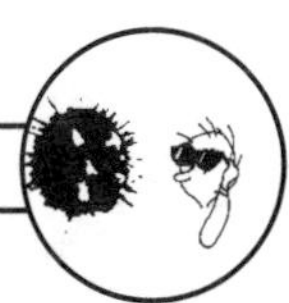

The Babu Bounce—pros and cons...discuss

Aedes aegypti is a peridomestic, anthropophillious mosquito. Those big, fancy words mean she has a special relationship with humans. The *Aedes aegypti* relies on us to provide her with food and breeding grounds. The various things that we discard which then fill with water are the perfect place for her eggs, and our legs below the knee are her favorite restaurants.

Secretaries in India constantly bounce their legs up and down—"the Babu Bounce"—to scare off the nervous beastie. Ironically, this makes the *Aedes aegypti* even more deadly. Because of the constant interruptions, she will end up poking you several times before getting what she came for. And every time she stabs you she can leave behind pathogens, whether she draws blood or not.*

**Mosquitoes...there's just no winning with the little bloodsuckers.*

Stop yer itchin'!

Try these remedies to halt the allergic reaction that makes mosquito bites itch*:

- Above all, don't scratch: It moves the saliva around and makes it itch worse.
- In the early part of the twentieth century, lemon juice, vinegar, oil of peppermint, tea tree oil, and oil of pennyroyal were used to help stop the itch.
- Rubbing alcohol effectively kills the sting but causes considerable pain.
- Icing the bump will reduce inflammation and swelling.
- Epsom salts and hot water work well together. One tablespoon of the salts dissolved in one quart of hot water, chilled, and applied to the bite, will help.
- Put the hottest water you can stand on a cloth and slap it on the bite. The itching should intensify briefly and then stop.
- Try putting toothpaste on the bite (must be a white paste, not a gel).
- Make a paste of baking soda and water or meat tenderizer and water and place it on the bite.
- Dab a strong infusion of the aromatic herb chervilon the affected area at regular intervals.

*Will making an "X" over the bite with your fingernails stop the itch? No! It's a myth!

Stop yer itchin', the sequel!

You've been bitten again. So what do you do, besides slapping the little beastie into a bloody smear? According to herbal medicine expert Susun Weed, there are many natural herbal remedies for insect bites.*

Plantain, a common weed also known as pig's ear or ribwort, can be recognized by the five parallel veins in its leaf. Simply chew up a leaf and place it on the bite to alleviate pain and swelling. Other leaves can be used in much the same way. Weed lists comfrey, yellow dock, wild geranium, wild mallow, chickweed, yarrow, and maple.

She cautions (and we emphasize) that you should be sure to correctly identify plants before chewing on them and to immediately spit out anything that tastes really bad or hurts your mouth.

**Yes, she goes by the name of "Weed" and she's into plants. We didn't make this up.*

Mosquito Miscellany

Tall tales and other tidbits

How Mosquitoes Came to Be*

Long ago there was a giant who loved to kill humans, eat their flesh, and drink their blood. "Unless we get rid of this giant," people said, "none of us will be left." One man said, "I think I know how to kill the monster."

He went to the place where the giant had last been seen. There he lay down and pretended to be dead. Soon the giant came along. He touched the body and said, "Ah, good, this one is still warm and fresh."

The giant flung the man over his shoulder and carried him home. He dropped the man near the fireplace and went to get some firewood.

As soon as the giant had left, the man got up and grabbed the giant's knife. Just then the giant's son came in. He was still small as giants go, and the man held the big knife to his throat. "Where's your father's heart?" The giant's son was scared. He said, "My father's heart is in his left heel." Just then the giant's left foot appeared in the entrance. The man plunged the knife

**Paraphrased from a Tlingit Indian legend.*

into the heel. The giant screamed and fell down dead, yet he still spoke. "Though you killed me, I'm going to keep on eating you and all the other humans in the world forever!"

"That's what you think!" said the man. He cut the giant's body into pieces and burned each one in the fire. Then he took the ashes and threw them into the air for the winds to scatter.

Instantly each of the particles turned into a mosquito. The ashes became a cloud of mosquitoes, and from their midst the man heard the giant's voice laughing, saying, "Yes, I'll eat you people until the end of time." The man felt a sting, and a mosquito started sucking his blood, and then many mosquitoes stung him, and he began to scratch himself.*

**The question remains: whatever happened to the giant's son?*

Another mosquito legend*

Long, long ago, there was an old woman who slept inside a house near a big village. She kept her door shut all of the time. The village boys wanted to know what was inside, but she would tell no one.

One day, some bad men stole bananas from the old woman's garden. When she saw that her bananas were stolen, she got terribly angry and stormed into the village, demanding to know who stole her bananas. The men denied knowing anything about her stolen bananas. So she threatened them: "If you don't tell me who stole my bananas I'll imprison you. I'll open the door of my house," said the old woman.

The men just laughed at the idea of being imprisoned by an open door, but the boys, hearing this conversation, realized that now might be the time they found out about what was in the old woman's house. After all, she was about to open the door.

This one's from the East Sepik Province of Papua New Guinea.

So all the boys rushed to the old woman's house to find out what was inside. And yes, they did find out. As soon as she opened the door thousands of mosquitoes came flying out of the house.

The boys had never seen mosquitoes before and just wondered what they were. They found out soon enough, as the mosquitoes swarmed them and started biting them and sucking their blood. The boys sprinted back to the village and the mosquitoes followed. Soon, the entire village was hopping with mosquitoes.

The people in the village all repented for how they had treated the old woman. "We didn't want to help the old woman find her bananas, and now she has unleashed the mosquitoes on us. She was a very good woman, and for all this time had kept the mosquitoes from us. We should have helped her!" the people of the village cried. But by now it was too late, the mosquitoes were out, and continue to fly about.*

*Thanks to Pan Asia Networking. More about Sepik stories at www.panasia.org.sg/sepik/html.

The Lion and the Mosquito*

Once upon a time a tiny mosquito started to buzz 'round a lion he met. "Go away!" grumbled the sleepy lion, smacking his own cheek in an attempt to drive the insect away. "Why should I?" demanded the mosquito. "You're king of the jungle, not of the air! I'll fly wherever I want and land wherever I please." And so saying, he tickled the lion's ear.

In the hope of crushing the insect, the lion boxed his own ears, but the mosquito slipped away from the now dazed lion. "I don't feel it any more. Either it's squashed or it's gone away."

But at that very moment, the irritating buzz began again, and the mosquito flew into the lion's nose. Wild with rage, the lion leapt to his hind legs and started to rain punches on his own nose. But the insect, safe inside, refused to budge. With a swollen nose and watery eyes, the lion gave a terrific sneeze, blasting the mosquito out.

Angry at being dislodged so abruptly, the mosquito returned to the attack: Buzz! Buzz! It whizzed round the lion's head. Large and tough as the lion was, he could not rid himself of his

**A very grim fairy tale from the Brothers Grimm.*

tiny tormentor. This made him angrier still, and he roared fiercely. At the sound of his terrible voice, all the forest creatures fled in fear; but, paying no heed to the exhausted lion, the mosquito said triumphantly, "There you are, king of the jungle! Foiled by a tiny mosquito like me!"

And highly delighted with his victory, off he buzzed. But he did not notice a spider's web hanging close by, and soon he was turning and twisting, trying to escape from the trap set by a large spider. "Bah!" said the spider in disgust, as he ate it. "Another tiny mosquito. Not much to get excited about, but better than nothing. I was hoping for something more substantial...."

And that's what became of the mosquito that foiled the lion!*

**The spider was eaten by a bird, who in turn was eaten by a cat; eventually a house was built.*

The Fox and the Mosquitoes*

A fox, after crossing a river, got its tail entangled in a bush, and could not move. A number of mosquitoes, seeing its plight, settled upon it and enjoyed a good meal undisturbed by its tail. A hedgehog strolling by took pity upon the fox and went up to him:

"You are in a bad way, neighbor," said the hedgehog. "Shall I relieve you by driving off those mosquitoes who are sucking your blood?"

"Thank you, Master Hedgehog," said the fox. "But I would rather not."

"Why, how is that?" asked the hedgehog.

"Well, you see," was the answer, "these mosquitoes have had their fill; if you drive these away, others will come with fresh appetite and bleed me to death."

One of Aesop's fables.

Swat teams

The Wonderlake Campground in Denali National Park, Alaska, holds an annual contest to see who can slap the most mosquitoes. Not to be outdone, Pelkosenniemi, Finland, has played home to an annual "World Championship of Mosquito Killing" since 1992. Teams of anti-mosquito slappers swarm the mosquito-thick highlands surrounding the town while being cheered on by throngs of spectators. The record? A mere seven kills in five minutes. Organizers blame the small number of kills on the very crowd the event gathers.*

**Mosquitoes are drawn away from competitors by the smell of the crowd (and the roar of the greasepaint).*

Sjaunja Myggmuseum

In 1998 local school administrator Ulf Lundberg had an idea to make the town of Gällivare, Sweden, a few extra Euros. Gällivare sits near the most mosquito-dense area of Sweden (3,300 mosquitoes per square meter) along Europe's largest marsh, *Sjaunjamyren* (Sjaunja marsh). Lundberg thought folks taking a break as the train stopped in Gällivare could spend time (and money) in a mosquito museum.

The museum features mosquito biology, tips on how to protect against mosquitoes, and entertaining mosquito facts.* Visitors view mosquito-related "interesting objects" as they are led by mosquito guides, "older people adept to living in the wilderness." Souvenirs are available, including handcrafted mosquito backpacks and locally manufactured tin mosquitoes. Visit online at www.gellivare.se/turism.

**Sound like a great idea for a book!*

Mosquito class

According to *Backpacker* magazine, Teaching Drum Outdoor School in Wisconsin teaches a mosquito course in which participants bare themselves "to Sister Mosquito in order to clearly hear her voice." The course teaches students to revere the mosquito as a respected guardian of the North Woods. "In the Hoop of Life," says instructor Tamarack Song, "Mosquito is as vital and noble and beautiful as the Hawk or Grandfather Pine." Experienced campers conclude that the best ways to avoid Sister Mosquito is to "stay high and dry, avoid activity at dusk and dawn, stay out of the shadows and in the breeze, go slowly, wear green, and be first in line [during walks]."*

**Oh, and one more thing: "Don't breathe."*

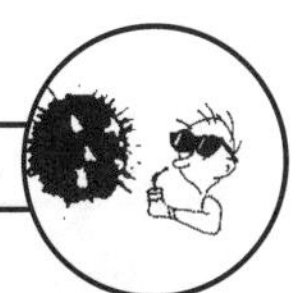

Now that's what we call camping

Entomologist C.D. Steelman relates a story of an insect-collecting trip accompanied by an entomologist friend:

[W]e saw a large, dark cloud...[and] thought the marsh was on fire...[b]ut we soon realized that we were seeing a vast swarm of countless millions of salt marsh mosquitoes.... Since the evening breeze was blowing the mosquitoes towards us, we quickly smeared ourselves with insect repellent.... Shortly thereafter [h]undreds hovered just in front of our faces; some bit despite the repellent, and in the brief period during which we removed our boots before taking refuge in our screened tent, scores of them landed on our repellent-free ankles and tortured us with their bites.... In the Texas coastal marsh, cattle-raising is uneconomical because mosquito attack reduces feeding time while the cattle are fighting mosquitoes.*

**These massive mosquito swarms have even been known to kill young and weak cattle.*

The great sheep and bunny war

Thomas Austin wanted to hunt rabbits on his estate in Australia.* So he brought just twenty-four wild European rabbits there from England in 1859. By 1951, the rabbits were overrunning sheep-herding land in Australia, eating enough grass to feed 40 million sheep! Sheepherders tried introducing hawks, weasels, snakes, and thousands of miles of rabbit-proof fencing, but that didn't stop these bunnies, who kept eating and eating and eating. Finally, the government introduced myxoma, a mosquito-borne virus from South America. European rabbits had no resistance to the disease, which was nearly always fatal. It took only three years for the mosquitoes to carry myxomatosis** throughout most of the bunny population, lowering it enough—in some areas, ninety-nine percent of the rabbits were eliminated—that foxes and dingoes could take care of the rest.

"Be vewy quiet, mate. I'm hunting wabbits." ***Also a song off Radiohead's* Hail to the Thief.

Mosquitoes and racism

As previously noted, when Europeans and Americans moved into new regions of the world, they had no immunities to local strains of mosquito-borne diseases, and would consequently be hit much harder than the native population. Ronald Ross, who had proven mosquitoes carry malaria, wrote that the disease "strikes down not only the indigenous barbaric population but, with greater certainty, the pioneers of civilization—the planter, the trader, the missionary, and the soldier. It is therefore the principal and gigantic ally of Barbarism." Malaria was seen as part of the so-called "white man's burden," a price to be paid as western culture was spread, often forcibly, around the world.*

**In these more enlightened days, we recognize that the white man's burden is actually monster trucks, yodeling, and clog dancing.*

What's the "Mosquito Coast"?

It's a region of eastern Nicaragua and northeast Honduras. It was a British colony from 1655 until 1860, when it became the Mosquito Kingdom. Nicaragua took over in 1894, and Honduras has controlled its share since 1960.*

*Mosquito Coast *is also the title of a movie starring Harrison Ford, based on a novel of the same name by author Paul Theroux.*

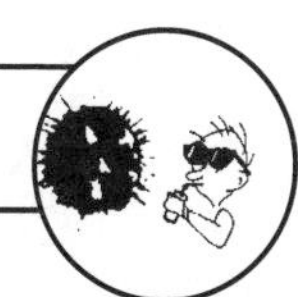

Planet of the Mosquito Movies

The first skeeter flick was Winsor McCay's brilliant 1912 black-and-white silent animated *How a Mosquito Operates*, in which a mosquito dines on the blood of a sleeping drunk and suffers the consequences. The next animated mosquito silent was 1920's *The Thrilling Drill.* A mosquito, frustrated in its attempts to bite humans, attends a meeting of Bolshevik mosquitoes and joins in yelling radical slogans: "Down with everybody and everything!"

Beneath the Planet of the Mosquito Movies

Hollywood loves giant bugs! Over the years, screens have featured huge tarantulas, ants, preying mantis (manti?), grasshoppers, scorpions, and wasps. In 1993,mosquitoes got their due with *Skeeter*, a tale of mosquitoes mutated by toxic waste. The movie had two taglines: "Earth is the final breeding ground" and "An environmental disaster with a name." Apparently the producers didn't notice the second tagline sounded like it referred to the movie itself.*

**It was a big year for giant mutated blood-sucking insects; 1993 also saw the release of* Ticks.

Escape from the Planet of the Mosquito Movies

The idea of giant mosquitoes was just too good for only one movie, so in 1995 came *Mosquito* (aka *Blood Fever* and *Nightswarm*). The *Blockbuster Entertainment Guide* describes the movie as an "Above-average giant bugs-on-the loose flick [that] pits campers and escaped convicts against human-sized skeeters that had mutated after drinking the blood of crash-landed aliens."

Conquest of the Planet of the Mosquito Movies

In 1999 came the German short film *Mosquito*, certainly the best film ever made featuring a love triangle between a man, a woman, and a bug.*

Battle for the Planet of the Mosquito Movies

In the 2002 Austrian comedy short *Mosquito Night*, an elusive mosquito drives our sleepless hero crazy. But is it really just a mosquito? Or does the fate of Earth itself hang in the balance?

**Depending on how you classify Mickey Rourke.*

Useful skeeter trivia!

In *Jurassic Park*, scientists cloned dinosaurs using dino DNA found inside ancient mosquitoes trapped in amber. Could it work?

Not in our lifetime. The DNA will probably not be complete (as in the movie, where scientists supplemented the DNA with frog DNA), so it would be awfully tough to get it right. Even if they could find complete DNA, the dinosaur egg would still need a place to develop. You see, even cloned sheep need real live sheep to grow inside; the scientists would need a real live dinosaur in which the eggs could develop.

Useless skeeter trivia!

In our favorite episode of *Gilligan's Island*, a Beatle-esque pop outfit shows up to get away from their rabid fans. What was the band's name?

The Mosquitoes!*

Unfortunately, the Mosquitoes were repelled off the island without saving our beloved castaways for fear that the Honey Bees (Mary Ann, Ginger, and Mrs. Howell) were better performers than they were.

D'ohp! Gilligan!

**Not John, Paul, George, and Ringo but Bingo, Bango, Bongo, and Irving!*

Bibliography: Where we found our facts

Adams, Sean. "A High-Tech Mosquito Barrier." *Agricultural Research* (March 1996): 12-15.

Adler, Tina. "Mauling Mosquitoes Naturally." *Science News* 91 (1996): 270–272.

Aldhous, Peter. "Malaria: Focus on Mosquito Genes." *Science* 261 (1993): 546–548.

"All the Buzz." *Sports Illustrated*, 7/24/95, p. 14.

Angier, Natalie. "For Monkeys, a Millipede a Day Keeps Mosquitoes Away." *The New York Times*, December 5, 2000.

Annals of American Medicine. http://www.annals.org/cgi/content/full/128/11/931, 12/03/03.

Barr, Ralph H. *The Mosquitoes of Minnesota.* Minneapolis: U of MN Experimental Station (Technical Bulletin 228), 1958.

Bates, Marston. *The Natural History of Mosquitoes.* New York: MacMillan, 1949.

Berenbaum, Mary. "The Natives Knew." *ChemTech* (May 1990): 275–279.

Bradshaw, William, and Christina Holzapfel. "Life in a Deathtrap." Natural History (July 1991): 35–36.

Carpenter, Stanley J., and Walter J. LaCrosse. *Mosquitoes of North America*. Berkeley, CA: University of California Press,1955.

Center for Disease Control. "Overview of West Nile Virus." www.cdc.gov/ncidod/dvbid/westnile/, 2003.

——. "Symptoms of West Nile Virus, Updated." www.cdc.gov/ncidod/dvbid/westnile/, 2003.

——. "West Nile Virus and Dogs and Cats." www.cdc.gov/ncidod/dvbid/westnile/, 2003.

Child, Elizabeth. "Zen and the Art of Living with Mosquitoes." *Skyway News*, 5/15-21/97, p. 9.

Clancy, Frank. "The Mosquito Fighter's Survival Guide." *Health* 9, no. 4 (1995): 84.

Clark County Mosquito Control District. "A Brief History of Control Methods." http://pithaya9.com/mosquito/control.html, 2000.

Day, Jonathan F. "Epidemic Proportions." *Natural History* (July 1991): 50–53.

Dinn, Michael. "A Little of the History that Colours Newfoundland." Newfoundlanders Read The Shipping News, www.educ.mun.ca/educ4142/.

Edman, John D. "Biting the Hand That Feeds You." *Natural History* (July 1991): 8–10.

Erdoes, Richard, and Alfonso Ortiz. *American Indian Myths and Legends*. New York: Pantheon Books, 1984.

Frugate, Craig. "Be Mosquito Smart: Tips To Preventing Mosquito-borne Diseases." www.disastersrus.org.

Gillette, Becky. "Controlling Mosquitoes Biologically." *BioScience* 38, no. 2 (1988): 80–81.

Gillette, J.D. *The Mosquito*. New York: Doubleday and Company, 1972.

Grey Bruce Health Unit. "West Nile Fact Sheet." http://www.publichealthgreybruce.on.ca/WestNileVirus/FactSheets/NonDeetRepellents.htm, 12/03/03.

Harwood, Robert F., and Maurice T. James. *Entomology in Human and Animal Health*, 7th Edition. New York: MacMillan, 1979.

Hailey, J. "Mosquito synopsis." imdb.com.

Hawley, William A. "Adaptable Immigrant." Natural History (July 1991): 55–59.

Herms, W.B., and H.F. Gray. *Mosquito Control*. New York: The Commonwealth Fund, 1994.

Hertzberg, Hendrik. "Summer's Bloodsuckers." *Time*, 8/10/92, p. 46–48.

Hull, Dr. Janet. "How to Beat those Summer Skeeters." SweetPoison Newsletter (May 2003): www.sweetpoison.com.

The Internet Movie Database, imdb.com.

Klowden, Marc J. "Blood, Sex, and the Mosquito." *BioScience* 45, no. 5 (1995): 326–331.

——. "Tales of a Mosquito Psychologist." *Natural History* (July 1991): 48–50.

Lauerman, John F. "Mosquito with a Mission." *Health* (March 1991): 58–60.

Levin, Ted. "The Mosquito Coast." *Sports Illustrated*, 6/24/96, p. 5-8.

The Mad Phat Internet Mini-Mall, madphat.com.

Mattingly, P. F. *The Biology of Mosquito-Borne Disease*. New York: American Elsener Publishing Co., 1969.

McCall, Bruce. "Five Historical Cigars." *Forbes*. 10/23/95, p. 182-183.

Miller, Henry I. "Is There a Place for DDT?" *The New York Times*, August 7, 2003.

Mosquito.org. "Mosquito Traps." http://www.mosquito.org/MosqInfo/Traps.htm, 12/09/03.

Nagel, Ronald L. "Malaria's Genetic Billiards Game." *Natural History* (July 1991): 59–61.

Nielson, Lewis T. "Mosquitoes Unlimited." *Natural History* (July 1991): 4–6.

North Carolina State University. "Urban Reppellents." http://www.ces.ncsu.edu/depts/ent/notes/Urban/repel.htm, 12/04/03.

Pan Asia Networking. "Sepik Stories." http://www.panasia.org.sg/sepik/html/nat03.htm, 12/08/03.

Perth District Health Unit. http://www.pdhu.on.ca/lifestyle/fightbit.htm, 12/03/03.

Rennicke, Jeff. "Song of the Bloodsuckers." *Backpacker Magazine* (April 1997).

Revkin, Andrew C. "West Nile Moving Faster and Wider." *The New York Times*, August 8, 2003.

Spielman, Andrew and Michael D'Antonio. *Mosquito: The Story of Man's Deadliest Foe*. New York: Hyperion, 2001.

Springer, Ilene. "Home Remedies that Really Work." *Ladies' Home Journal* (June 1993): 80.

Steel, Scott. "Summer's Sting." *MacLean's*, 6/26/95, p. 46–47.

Trager, James. *The People's Chronology*. New York: Henry Holt and Company, Inc., 1996.

United States Department of Agriculture Animal and Plant Health Inspection Service. "2003 Equine WNV Outlook for the United States." APHIS Info Sheet, June 2003.

Waldbauer, Gilbert. *What Good Are Bugs? Insects In The Web Of Life*. Cambridge, Massachusetts: Harvard University Press, 2003.

Weed, Susun S. "Ease Those Bug Bites with Easy Herbs." www.susunweed.com, 2000.

The Why Files. "Climate Change = More Disease?" Mosquito Bytes: http://whyfiles.org/016skeeter/6.html.

Wilson, Samuel M. "Pandora's Bite." *Natural History* (July 1991): 26–29.

Index: A fast way to find our facts

Aesop's Fables:
- "The Fox and the M" 134

Amistad slave revolt **67**
Anopheles Ann **61**
Australia **139**
Attraction to humans **45–48**
"Babu Bounce, The" **124**
Babylon **56**
Benefits of **54**
Beothuk tribe **121**
"Bites"
- anticoagulant 8, 49, 51
- blood withdrawal 7–9, 11
- frequency of bites 12, 42
- human ankles 48
- immunity 50, 52
- itching remedies 125, 126
- poisonous? 51
- reaction to bites 49–51

Bonaparte, Napoleon **66**
Botflies and **10**
Cabot, Captain John **121**
China, ancient **56**
Cigar of Death, Black **94**
Clothing to avoid **123**
Columbus **56**
Control **83–100**
- Altosid 93
- bats 90
- Bti 91
- crude oil 84
- DDT 89
- dragonflies 86
- genetic engineering 97, 98
- goldenbear oil 93
- home & yard control 100
- Malathion 92
- mosquito fish 87
- naled 92
- pennyroyal, oil of 84
- purple martins 88
- Scourge 93
- sterilization 99
- traps 95, 96

Crane flies **2**
Daddy-long-legs **2**
D'Antonio, Michael **4**
Denali National Park **135**
De Soto, Hernando **60**
Disease **55–82**
- AIDS (HIV Virus) 82
- dengue 76, 77
- diagnosis, use in 54
- Ebola 82
- encephalitis 78–79
- filariasis 75, 77
- heartworm 81
- malaria .... 56–59, 61–63, 71, 73, ... 74, 85, 89, 140
- myxomatosis 139
- West Nile Virus 77, 79, 80
- yellow fever 59, 64–70, 72–74, 77

Eiffel, Gustave **73**
Etymology (word origin) **34**
Evasion of **123**
Feeding
- amount of blood taken per bite 11
- attraction 45–48

Feeding, continued...
- blood ingestion limitations 16, 17
- blood withdrawal....................7–9, 11
- cannibalism....................................15
- frequency of blood meals 12
- times of blood meals 31
- non-blood feeding 3
- prey 14
- use of blood 4

Flight Range..................................29
Flight Speed..................................29
Finlay, Carlos Juan.........................74
Foreign names for..........................35
Genghis Khan.................................56
"Gilligan's Island"........................144
Gorgas, W. C. 72–74
Grimm's fairy tale:
- "The Lion and the Mosquito" 132

Gynandromorph mosquitoes...........6
Hawaii 63
Hispaniola slave rebellion 65, 66
Itching remedies.................125–126
infections.......................................59
Jefferson, Thomas..........................66
Jurassic Park...............................144
Laveran, Dr. Charles 71, 74
Lesseps, Ferdinand and Charles....73
Louisiana Purchase 66
MacArthur, General Douglas.........61
Malaria Moe 61
Manson, Sir Patrick 74
Memphis 70, 71
Monkeys...................................64, 111
Mosquito, defined............................2
Mosquito class 137
Mosquito Coast............................141
Movies about mosquitoes ...142–143
Mudd, Dr. Samuel 69
Nematodes.....................................75
New Jersey 85
New Orleans 66, 68
Origin myths........................128–131
Nonhuman victims 14, 53, 79, 81
Panama Canal..........................73, 74
Pelkosenniemi, Finland................135
Philadelphia............................65, 68
Physiology
- brain size 36, 37
- buzzing 43
- life span 32
- male sex organs 21, 22
- sense of smell 45–48
- weight..38

Predators of 39
Racism and 68, 140
Reed, Walter 72, 74
Repellents 101-126
- Avon's Skin-So-Soft 105
- Bounce fabric softener 119
- citronella 106–108
- citrosa 108
- chrysanthemums 109
- coils 114
- DEET...................................102–104
- foggers 115
- geraniums.....................................109
- garlic.....................................106, 110

Repellents, continued...

grass of China 107
millipedes..................................111
mud. 120
Native American repellents..........122
permethrin..........................112–113
pyrethrum 112
pyrethroid...........................114, 115
smoke 118, 123
sonic devices 117
spicy foods 110
vitamin B1 110
zappers 116

Reproduction
attraction to mate 19, 20
blood's role.....................................4
"castration" 22
decapitation 22
egg laying 12, 23–26
mating 19, 20, 24, 99

Rome 58

Ross, Sir Ronald 74, 140

Skeeter hawk 86

Slaves and 59, 64–68

Smith, John B. 85

Sociology
daily life 31
family life 27
hybernation 30
migration 28–30
population 40

Spanish-American War 72

Species 40
Aedes 25, 32, 106
aegypti 74, 97, 124
Anopheles 25, 32
culicifaces 25
Asian tiger mosquitoes 28
Cattail Mosquito............................26
Culex 25, 32
Floodwater mosquitoes 25
Pitcher Plant Mosquito 26

Spielman, Andrew 4

Steelman, C. D 138

Swatting
contests 135
difficulty 44

Toussaint-Louverture, François 66

Vietnam War 62

Washington, George 65

"White man's burden"................140

"White man's grave" 59

World War II 61

About the Authors

Scott D. Anderson was an an author, scholar, entrepreneur, and pilot from Duluth, Minnesota. Besides *The Mosquito Book*, Scott wrote *Unknown Rider* and *Distant Fires*, winner of the 1991 American Library Association Best Book for Young Adults. He also flew F-16s for the Minnesota Air National Guard. Scott died in Duluth in 2000 after crash landing while test piloting a plane. *The Mosquito Book* was his idea.

Tony Dierckins operates X-communication, a publishing house in Duluth, Minnesota. He has coauthored over a dozen books, including *True North: Alternative and Off-Beat Destinations in and Around Duluth, Superior, and the Shores of Lake Superior, Greetings from Duluth* Volumes I and II, *The WD-40 Book*, and the *Duct Tape* books and calendars. He also created the CoasterBook series published by Running Press.

Scott Pearson lives in St. Paul, Minnesota, with his wife Sandra and daughter Ella. He is a writer and copyeditor who has contributed to every book published by X-communication. Scott's story "The Mailbox" won a 1987 *Minnesota Monthly* Tamarack Award. Since then he has published humor and poetry in regional anthologies and periodicals. His short story "Full Circle" appears in the 2004 *Star Trek* anthology *Strange New Worlds VII*.

Bonus: Use this book to repel and exterminate mosquitoes!

When you've finished reading *The Mosquito Book*, put it to good use in your battle against mosquitoes! By lighting this book on fire, you will produce smoke that will keep mosquitoes away from your immediate area as long as it is burning. If you're against burning books, take an old wooden spoon from the kitchen drawer, duct tape it to this book, and use your new tool to slap at skeeters while keeping your hands blood free!